Mark Wheeller

Game Over

*A Verbatim play using the words of
Breck Bednar's family and friends*

*Commissioned by Beaumont School with the The National
Lottery Community Fund.*

Salamander Street

PLAYS

Game Over published by Zinc Communicate in 2019
ISBN 978-1-902843 44 5

This edition first published in 2020 by Salamander Street Ltd.,
272 Bath Street, Glasgow, G2 4JR (info@salamanderstreet.com)

Reprinted in 2021

Game Over © Mark Wheeller, 2019

All rights reserved.

All rights whatsoever in this play are strictly reserved and application for performance etc. should be made before rehearsal to MBA Literary Agents Ltd, 62 Grafton Way, London W1T 5DW (attn: Sophie Gorell Barnes).
No performance may be given unless a license has been obtained.

You may not copy, store, distribute, transmit, reproduce or otherwise make available this publication (or any part of it) in any form, or binding or by any means (print, electronic, digital, optical, mechanical, photocopying, recording or otherwise), without the prior written permission of the publisher. Any person who does any unauthorised act in relation to this publication may be liable to criminal prosecution and civil claims for damages.

PB ISBN: 9781913630263
E ISBN: 9781913630270

Cover and text design by Konstantinos Vasdekis

Printed and bound in Great Britain

10 9 8 7 6 5 4 3 2

Further copies of this publication can be purchased from
www.salamanderstreet.com

CONTENTS

Introduction v

A Note from Lorin LaFave xii

Introduction to Directing Game Over xvii

Introduction to Acting in Game Over xx

Game Over 1

Acknowledgements

Lorin LaFave, Barry, Chloe, Carly & Sebastian Bednar for their words and permission to use this emotive story.

Ollie and Matt, Breck's friends. Thank you for (along with the triplets) giving the play the all-important teenage relatable factor.

Rachael, and my family, who encourage my commitment to writing these plays.

Lynsey Wallace & Zoë Shepherd from Beaumont School, St Albans in Hertfordshire who gave me the opportunity to create this important play for Breck's family.

The production team at Beaumont School for their work on the premiere.

Lynda Taylor from Zinc Communicate who, along with the team at dbda (Zinc's previous incarnation) published my plays when no one else would.

Thanks to George Spender and those in the Salamander Street team for their efforts to extend the reach of my plays.

Sophie Gorell Barnes and all at MBA for their continued support and belief.

Introduction

I am often asked how I arrive at the subject matter for my plays.

For most of my career I was a teacher/youth theatre director, so would choose subjects to interest/motivate my groups to produce their best work. The skills these performers picked up would cascade into the drama classroom. I guess that's a factor in whatever I do but, as time has moved on, I was offered commissions where I had no choice about the content. This explains why I wrote so many road safety plays, commissioned off the back of the surprising success of *Too Much Punch for Judy*, rather than any passion I had about road safety.

In recent years, I have been fortunate enough to be commissioned and allowed to choose the content I think will engage and motivate young people in their drama/theatre work.

The story of how I came to author *Game Over* significantly links to a childhood experience of my own... more of that later.

Game Over began with Fiona Spargo-Mabbs, mother of Dan, (from my 2016 Methuen Drama play *I Love You Mum, I Promise I Won't Die*) and a Facebook post she made when she visited 10 Downing Street. Fiona was delivering a letter on behalf of the PSHE Association to campaign for statutory PSHE in schools. Alongside her were two mothers; Sasha Langton-Gilkes who lost her son to a brain tumour and Breck's mum, Lorin. I had seen nothing of Breck's story in the media despite extensive coverage. As soon as I saw Fiona's post I looked Breck's story up and remember thinking...

"I'd love to write a play about that!"

I spoke to Fiona, and she revealed Lorin had asked her about *I Love You Mum* and had expressed interest in a play telling Breck's story. Fiona put us in contact but, given that I had just retired from running a youth theatre group, warned Lorin that it would need to be commissioned and a performance group found.

Cut to October 2017 and out of the blue I receive an email from Zoë Shepherd, Assistant Headteacher, at Beaumont School, in St Albans, Hertfordshire.

Dear Mark,

We have just been awarded Artsmark Platinum and are keen to keep driving things forward. We have been looking for a play suitable for our Year 9 Drama Company to perform that explores LGBT+ issues. Anything we read doesn't seem to be what we need. As we were discussing it, I said it would be great if there was a Mark Wheeller play (we study Missing Dan Nolan *as our OCR GCSE set text).*

Is there anything in the pipeline or is this something you would be interested in exploring with our school if we could find funding for a commission?
Zoë.

Here's another coincidence! Two months previously, I had agreed to dramatise a transgender story for another school. Funding was, I was told, being sorted. Contracts were drawn up and I had signed my side of them. I didn't want to write two similar plays so I explained the situation and cheekily offered up the possibility of Breck's story. (The transgender play never happened, as the commissioning school were unable to achieve the funds needed to develop a play – it's not easy.)

Meanwhile, Beaumont had jumped at the idea of using Breck's story and made a successful bid to the National Lottery Community Fund for the project… writing and production.

Things happened fast! Lynsey Wallace, Head of Drama at Beaumont and responsible for organising the bid was incredibly proactive, giving me total confidence in her ability to make this whole project work.

The bid was granted in August 2018.

Between September and October various permissions were sought, contracts signed and, throughout October, I conducted the interviews.

Those with Lorin and the triplets were conducted in their home in three segments across about six hours. I spent the afternoon with Lorin alone, who told me everything up until the Spanish holiday.

The triplets arrived home from school and told their version of the whole story, with Lorin chipping in. It had been unclear whether they would be willing to talk so it was such a relief to me that they did. The teenage voice is always a benefit to any play where teenagers will be reading/performing it.

Their version of events, reflecting on an incident that happened when they were in Year 7 from the vantage point of now being in Year 12, was fascinating and hugely emotive (as you will see when you read the play).

Some aspects of what they said (for example looking at information about the murder online) had never been discussed in front of their mum. I felt honoured to be a part of this open exchange. Normally I interview "subjects" separately. This group interview was full of interaction, which proved helpful for the scripting of the play.

Lorin concluded the day's interviews by her telling me the dreadful details of the murder and her reaction to it. She must have been exhausted! I had a long drive home to process it all and knew (as if I didn't before) I had an incredible and important tale to tell. I couldn't wait to start the transcription process and did… the very next day!

Ollie and Matt (real names) offered their stories on behalf of Breck's friends. I conducted their interviews by phone… poor Ollie had to deal with my limited IT skills when I managed to wipe the recording as soon as he was off the phone. He generously gave me a second interview… and recorded that at the other end of the phone just in case!

Finally, I interviewed Barry, Breck's dad in his workplace. I was fascinated by his attitude to LD, and he requested that I make an effort to include this in the play. I knew how the verbatim genre would allow for this very easily.

I blitzed the transcription of all the interviews.

Lorin 25,556 words
Triplets and Lorin: 8,915 words
Matt 7,536 words
Ollie 4,313 (shorter possibly because it was a second take) words
Barry 8,376 words
Total 54,696 words.

There were a few other documents relating to Lewis Daynes' emails and the secret recording to add to the total, which amounted to around 500 words.

I had to reduce these 55,000 (approx.) words to a play I was contracted to deliver at about 8,000. It transpired to be just over 12,000. I wanted to tell the story with the detail I felt it deserved. With no particular (word) target in mind, I edited each interview to leave what I felt would prove useful for the play.

I knew where I wanted it to begin.

The triplets had talked about their thoughts on moving house and the way their schools had reacted to such a painful situation. I remember, during their interview, thinking, the play must begin with these "normal" events and then to launch into this horrifying context. I went with my gut feeling and pasted their words from this part of the edited interview to form the main body of this section and then edited them a little. I didn't need to add anyone else's accounts, because no one referred to these moments because they weren't there! Simple.

With the remainder of the play, I used the chronology of the story and took whoever said the main details of the story as the backbone, and introduced other accounts to it where they added a different perspective.

As an afterthought, I decided to use the transcript (edited) of Lewis Daynes' call reporting the murder, as a prologue to the play. It is the most horrifying start to any of my plays and lets the audience know instantly where this unstoppable juggernaut of a story is heading. I intercut the phone call with commentary from Breck's parents, offering a key moment for Barry to highlight his views that I mentioned earlier.

One other idea I had while interviewing Lorin was to divide her role into multiple Lorins. I remember her saying:

'I don't feel whole, missing parts and can't function properly. Torn apart.'

As she said this, I had the idea of dividing her part into six. This would have the advantages of:

1. her being able to talk/fight with herself,

2. illustrating how Breck might have felt (and certainly how LD perceived her to be), surrounding him with her opinions, love and frustration and

3. preventing one character hogging the stage with long monologues that may lack movement.

I made "Lorin" (without a number) the original Lorin from her time with Breck, prior to LD splitting her apart. With his arrival in her world, the other Lorins start to appear, becoming increasingly concerned... even paranoid about the situation as the number allocated to the character increases.

I am aware that directors may wish to re-allocate these numbers (as happened in Beaumont School's premiere production, where Lorin 5 & 6 have limited lines) and I am totally happy about that. Directors must make it work for them, but the Lorin's must interact with each other and Breck/LD and never become standing narrators.

This concept of the "torn apart Lorin" is, for me, the iconic stylistic feature of this play. I can't wait to see the different ways it is interpreted as I see more productions.

Lynsey Wallace, Beaumont School Head of Drama, is an experienced director and developed the premiere performance with no involvement from me. I was intrigued to see how they presented it, more so because this has only ever happened with three of my thirty plus plays, (*Legal Weapon*, *The Gate Escape* and *Chequered Flags to Chequered Futures*). Each of those productions exceeded my expectations, so I was optimistic!

I couldn't have been be happier with the public reaction to it. It was breathtaking. People even said I had "crafted it beautifully". Not sure I've ever previously had that kind of comment about my use of peoples words. My CSE Grade 2 English Literature didn't hold me back.

Right from the off... they added a beautiful little movement piece with the Lorins bringing small symbolic stones onto the stage, then the ensemble (who were so disciplined and focussed on and off stage), going into the official opening sequence – gaming down the years. The gaming theme was cleverly carried through the production with each transition conducted in a gemlike manner and succeeding to place the actors in their correct position seamlessly. Outstanding.

The six (there were actually only five) Lorins who completely achieved my aim of enabling a more interesting relationship with Breck who can be surrounded by "mother" and take him more by surprise, suddenly targeting at him from a different angle.

What I hadn't predicted... and this was utterly brilliant... was the relationship LD, the murderer, had with Breck's parents! It was sinister, for example calmly fiddling with Breck's father's (Barry) tie knot at one point, unseen by Barry. This built to a horrifying climatic scene where he (Matt Sims – what a theatrical performance) lifted her, Lorin (Beth Macrae – a totally sustained and believable performance) in the air at the very moment she learns Breck has been killed. It was just awful, so wrong, like this awful, heinous crime. The performers were exceptional. All the lead roles were taken on with clarity, pace and ability.

Some smaller roles sometimes stood out too. For example Jack Wathen as Sebastian (one of Breck's siblings) who barely has any lines, but was on stage often, had to deal with the problems of not speaking a lot but reacting. It was never awkward. Fantastic performance!

Ruari Spooner and Jamie Carrick as Ollie and Matt (Breck's gaming friends) were also exceptional. Luke Nixon brought a wonderful maturity (and exceptional commanding posture) to the thoughtful words Barry (Breck's dad) brings to the play.

The highlight though for me was seeing Millie Moore and Hannah Hunter (who played Breck's other two siblings excellently and so naturally) meeting, chatting and laughing with the Chole (who I had the honour of sitting next to for the most horrifying parts of the play – which had me on edge) and Carly Bednar (Breck's real siblings) who had just been in the audience worrying me about how they might react to seeing these scenes brought to life. These moments of these young people (and others) just being teenagers together, enjoying each other's company, offered for me the deeper possibilities of healing I believe this play has. It was simply "one of those moments"... and I was only half watching.

The same applied to the Lorins all chatting away and laughing with the real Lorin LaFave. I hadn't expected that... and how brave the real Lorin was to stand up as soon as the play ended and thank the cast!

I am aware that some have suggested this story may be too shocking to share. The word "safeguarding" has been proffered to defend this view. I don't agree and fear young people are in danger of being safeguarded to an extreme where they are denied the opportunity to witness the warnings this story offers.

The TV film, *Gail is Dead* (still available on YouTube) shocked me into never messing with drugs. Shock tactics don't work with everyone… but for many, they do. I am one of them.

A drama teacher recently contacted me about *Hard to Swallow*. She said, before purchasing a set of scripts (which she felt were a crucial tool), she had to speak to her Child Protection Officer to get permission to use the script. My reply to her was:

> *"Arguably, there are more child protection issues by avoiding these topics than there are by arming young people with the knowledge that comes from true stories such as these."*

Had Lorin's complaint about LD's grooming behaviour been followed up properly, Breck would still be alive. As I indicated earlier, this resonates with an experience I had as a ten-year-old boy. I was interviewed about it on BBC1's *Panorama: Scandal in the Church of England*, April 2019 alongside another boy at my school.

The programme revealed (I wasn't aware of it until then) that I would not have been abused (low level and once only fortunately, unlike my schoolfriend) had the authorities concerned (the church) followed up a complaint that had been made some months before. These situations leave a feeling of "something has to be done".

The play gives a clear idea of what effect it has had on Lorin, her family and Breck's friends. Their passion to speak out about the situation they were hurled into emphasises how we should make every effort to tell/hear these stories and learn from them.

I hope you will join me in using *Game Over* to help to convey the Breck Foundation's message as far and wide as possible. I am excited to see how this play can both generate outstandingly imaginative drama and help to empower young people to make safer choices for themselves online.

Mark Wheeller, 2020

A Note from Lorin LaFave

When I was introduced to Mark Wheeller and recommended that he write a verbatim play about my son Breck's story, I was very keen to learn more about how this would work. I had founded a charity called the Breck Foundation in 2014 after poor fourteen-year-old Breck was groomed by an online predator who ran the server in his gaming group with friends from school. Sadly, after a year of spending time online together, he was lured to his death at the predator's flat.

Since then I have been delivering presentations to pupils at schools and speaking at training sessions and conferences about Child Sexual Exploitation and grooming, as well as campaigning for changes in laws, policies and policing, so the idea of having 'help' to share awareness and understanding of the signs of grooming and exploitation online really appealed to me. One of our most important ethos within the charity is to teach others to talk about issues they may be affected by with their family and friends, and to look out for each other. So, what better way to get young people talking about these issues than to allow them to reach and teach each other through the creative outlet of drama at school.

Mark spent time with me, Breck's triplet siblings, their dad, as well as some of his friends to get a realistic version of what was happening in Breck's life at the time. Mark also used the predator's voice through police recordings that had been obtained in the investigation, so this is truly a well-rounded view of the story and what was being said to the gamers during the year-long grooming process.

Breck and all of his friends had of course sat through e-safety training sessions at school, but at the time these lessons were not delivered in a way that enabled the pupils to really listen, they felt patronised, and they weren't given real-life examples, so they did not take in these important life lessons when presented just as 'rules'. Teens think they know all and indeed do know so much, but important life experiences that account for so much they may not yet have, so we want Breck's life, and sadly in his case, 'death' lessons to really resonate because everyone will have a friend or school colleague who may be engaging with strangers online. We try to emphasise that everyone online is a stranger, and that is not to say that they are all bad, but that online, we are not able to make an educated or safe decision whether they are genuine and safe or not. We teach to never meet up in a 'private' place when we have only met online.

In Breck's story and many others, a stranger online may infiltrate a young person's life, befriend them, make them laugh, mentor them, encourage them, but then gradually isolate, change, manipulate and control them so that the person being groomed cannot see it themselves, which is why it is so important for young people to know these signs so that they can look out for each other online whether interacting on social media or on gaming platforms. Predators will find a shared interest to trick the young person into believing they are 'real' friends. Breck was a clever, confident and liked boy, the kind of teenager who never thought they could fall for something like this, and yet sadly he did. He adored gaming and computing, and this is how the predator built a relationship with him, through their shared interests.

Whilst many teachers may be great at educating, young people are inundated with safety messages and reminders, so by using this hard-hitting play as a resource, with a true story of a real boy, they can better engage their pupils to be aware of the dangers they may face online, to teach them to recognise the signs of grooming and report these to a trusted adult, they will be better educated and this will empower them to look after others as well. They can then better communicate concerns, thereby keeping themselves and others safe from harm online and off as grooming happens in all forms, to girls and boys, and for various adverse outcomes, such as Child Sexual Exploitation, County Lines gang activity, radicalisation or any harmful outcome.

Whilst predators can be any age, this predator was only eighteen years old, a teenager like the other boys, and they were not 'afraid' of other guys their age. Many people saw the changes in Breck's personality, and I had reported to schools and police, but because Breck did not present himself as a 'vulnerable' child, no one took the warning signs of grooming seriously. We all have vulnerabilities and predators will prey on this. I did my best as a parent, but without the skills, knowledge and help, my efforts still failed, and that is every parent's absolute worst nightmare.

The work that I do now is to try to help others, so that no other child will go through what Breck did, this tragic loss of life of a chilled, innately good, productive human being with potential. My eyes tear up even now as I write this, as no matter how much time goes by, I still hugely miss my lovely son Breck, and nothing can bring him back.

Use this play to teach in the engaging way it is written, with Lorin, me, who has been pulled to bits, and all the lovely characters who didn't deserve to be a part of this tragedy, with the laughter that comes too. And I thank Mark for writing *Game Over* beautifully and Lindsay Wallace of Beaumont School in St Albans for obtaining the funding to make this happen and guiding our first showing with her amazing, talented and warm cast. They were attentive and conscientious as they wanted to honour the teenage boy in the best way they could, by ensuring the performances were to the best of their ability – what better way to learn than through enjoyment.

Most young people today are living so much of their lives out online, so by embracing the lessons through performance, the messages will further resonate. Breck's story offers an engaging platform for opening honest discussions about online behaviours.

I hope you enjoy performing the play together and I hope you will remember my Breck for the awesome guy he was.

Break a leg x

Lorin LaFave (Breck's Mom)
June 2020

Breck Bednar

www.breckfoundation.org

Introduction to Directing *Game Over*

When the script landed in my inbox I couldn't quite believe it was actually happening. Tentatively opening the document, the realisation that this was somebody's real life story became hugely overwhelming. It was not just a play to be staged and acted; it was a truth, a reality and had a huge message that needed to be told. This story would teach young and old of the dangers of online gaming and how privileged I felt to be working alongside Mark and Lorin in helping to do this.

Fast forward to December, where the cast, who had auditioned for the production prior to seeing a script, seated in a circle, scripts in hands eyed each other with excitement and apprehension. The reading was intense, as they carefully read each line with meaning and disbelief. The silence after the reading was not predicted. They were shocked - at what they had read, that it had really happened. How were they going to do this justice?

Now, it was time to address this as a play. Director and actor hats on!

The hardest challenge when approaching this text is the fact that the words themselves are so powerful, there is almost cause to debate how practical the production should be.

As a result of this thought, and wanting to provide our young cast with opportunity, the decision was made to keep the cast on stage throughout. Acting as a voice of society or a visual representation of social media. Communicating the fact that this could happen to anyone, anywhere and at any time. There lies a challenge in what the ensemble then become? How should they be? Where should they be?

At this point, with an on-stage ensemble and a word-heavy text, the physical exploration with the cast began through finding moments; snatches of scenes or memories made from words taken from the read through or even specific lines of text, we started playing with making pictures or bringing these moments alive. Using physical theatre devices, music, a row of stools, a piece of gaming equipment as a symbol, moments started to build and ideas began to flow with how physical this piece could be and how the ensemble would be key to make this work as a convention.

The fractured Lorin characters also encouraged the physical style of the piece. It was key that Lorin played the main voice and was the most real version of the character, the other numbered Lorins seemed to represent all the emotions and elements that make a human. We found it important to not make this explicit in our interpretation as it was decided that humans are not that straight forward and by having a unified voice, a familiarity of them all being one through matching gestures and vocal intonation, it would be more successful but as the script plays through, their element of emotion becomes more clearly conveyed through the physical gesture that still remained familiar. Further into the rehearsal process, it felt right to always have the Lorins positioned on the stage, looking on to the action, representing the truth of how as a mother, Lorin was groomed too.

Finding the rhythm of the play was important to gain some form of rise and fall. After a simple blocking of the entire piece and seeing it run, it felt all one pace, this is because we start with tragedy and we end with tragedy. Transitions became key in providing some form of break or link from one section to the next. They needed to contrast to create the rhythm and lead appropriately into the following section, so gaming, holding screens, Miis, characters from games or playing out scenes from games were all explored as options. With lighting, sound and media becoming integral to this decision making also.

The set was inspired by the fact that Breck had three screens, as Lorin references in the play. This gave the idea for using multi-media possibilities and shadow play. LD in the shadows and always there, making Breck physically look smaller, and more vulnerable, through the lighting device.

It was felt that we were not going to play this with accent or to try to make it true to the people who the play is about. Having a local accent and a personal, young interpretation of the role reinforced the message of the play and that this could have happened to anyone, anywhere. Independently from the rehearsal room, the cast had looked on social media, watched interviews and read articles to gain as much knowledge as possible. This certainly impacted on their making the roles more believable, yet personal.

Sharing a run through with Lorin and Mark was exciting but scary. None of us knew how either would react to what we had created. We were delighted by their responses. Lorin encouraged for more attack with the role of LD, really showing the disruption he caused to them all. She also encouraged to find the happy moments in the family dynamics. She laughed with Breck and this needed to be part of their story in the play too. Great feedback to really help develop character and dynamic of the production.

The play is powerful, there is no debate there. It has a heavy message that must be communicated. It needs to be relevant to our audience for them to hear. It needs to have light and shade, and even show moments of happiness as there are always moments and memories that are happy. Audiences may respond to this production differently, but this is a good thing, as you will then know you have made an impact.

Here's to seeing many more different productions of *Game Over*!

Lynsey Wallace

Head of Drama, Beaumont School

June 2020

Introduction to Acting in *Game Over*

Game Over was a hugely interesting text to perform and read. I remember the first time reading it, we were all just in shock over how the events took place as we learnt the story for the first time. Since we were the first cast, we read lots of different versions of the script throughout the writing process, so this reading and re-reading of the story engraved it in all of our minds. When we had finished reading it, we had engaged with the story and been shocked at the events. What came from that was a collective drive to perform the story with the justice it deserved: to share The Breck Foundation's message the best we could.

I performed two different roles in two different sets of performances. In the first shows, in the summer of 2019, I played "Ollie" who was one of Breck's friends who he played with online. In the second set of shows, in March 2020, I played LD – Breck's murderer. These two different roles were so contrasting but interesting to play, as I essentially went from performing as a victim to performing as a predator.

Rehearsals could be tricky sometimes. Especially after performing it so many times, you can become dull to the message/story which affects the performance. This is a challenge for those performing *Game Over* in the future – to always remember what it is you are performing and how real it is as well. As it's a verbatim play, you have to be in the moment with it at all times. Which is true to any good performance, but especially a play with this type of style.

We were all reminded of how real this story was when we met Lorin for the first time. This play that we had been performing, these roles that had only existed in the rehearsal rooms of Beaumont School, suddenly turned to reality. It was slightly surreal. I remember we were rehearsing for the whole school play – *Grimm's Tales* – at the time, when Mrs Wallace, our Head of Drama, told us Lorin was coming in to have a chat and meet the cast. And we did. She came in and we sat in a circle. We introduced who we were and the roles we were playing, then just asked her questions and talked.

Afterwards, on the way home, one of my friends and fellow cast member started crying after being confronted with the reality of the whole story. This awakened me to the reality of it as well.

The first performance went well and I just remember how invested we were, which reflected in the audience. The audience came out in shock but also with a newly educated and engaged mindset on the reality of online grooming. Afterwards, on the final night, we met Breck's family. This hit home and it made me proud that we had worked so hard to share Breck's message and the family's message. To share their words in a form of art, so that this message can keep on being shared and not forgotten – so that people, especially young people, can learn from this story.

Ruari Spooner

Ruari Spooner was fifteen years old when he played the role of Ollie in the premiere performances in July 2019, and later played the role of LD in March 2020. Ruari has previously performed with Forced Entertainment and has now successfully secured a place with The National Youth Theatre Company, 2020.

Game Over was commissioned and premiered on July 2019 by Beaumont School, St Albans, Hertfordshire with funding from The National Lottery Community Fund at the Trestle Arts Base, with the following cast:

Barry	Luke Nixon
Lorin	Beth Macrae
Lorin 2	Lydia Jacobs
Lorin 3	Jessica Payne
Lorin 4	Molly Robbins
Lorin 5	Izzy Storey
Breck	Tom Fletcher
Chloe	Millie Moore
Carly	Hannah Hunter
Sebastian	Jack Wathen
Matt	Jamie Carrick
Ollie	Ruari Spooner
Will	Stefano Geymonat
Joe	Genevieve Morris
LD	Matthew Sims
Operator/Policeman	Olivia Tetlow
Police	Hannah McCorkindale Brown
Police Officer/Teacher	Harry Dean
Girl	Ruby Nicholls
Friend	Sahar Hodgson
Flight attendant	Celeste Jones
Director	Lynsey Wallace
Associate Director	Zoë Shepherd
Lighting and Set Designer	Trevor Wallace
Composer	Luke Bainbridge & Yannick Mayaud
Additional Sound	Josh Seddon

Characters

LORIN BEDNAR
Breck's mother

LORIN 2-6
Breck's mother[2]

BARRY BEDNAR
Breck's father

BRECK BEDNAR

CHLOE BEDNAR / CARLY BEDNAR / SEBASTIAN BEDNAR
Breck's triplet siblings

MATT / OLLIE / WILL / JOE
Breck's gamer friends

LD AKA LEWIS DAYNES

OPERATOR

FRIENDS

TEACHER

POLICE

POLICE OFFICER

GIRL

ANONYMOUS GAMER FRIEND

FLIGHT ATTENDANT

8M, 9F AND 7 M OR F. DOUBLING
/ MULTIROLLING IS POSSIBLE

---

[2] From Lorin's feelings of being "torn apart". For more details about this concept please see my introduction to the play.

Section 1

999 CALL

The stage is dressed with heart-shaped stones made visible in significant places. **LORIN** *stands by the stones and picks one up to hold it tenderly. Some distance behind her is* **LD**, *wearing a black T-shirt with no logo. The music/lighting indicates the entrance of an ominous presence. His face is partially concealed by a half mask. This remains throughout the play until the courtroom scene. He is subtly disrespectful to* **LORIN**. *His disrespect continues throughout the play and becomes more brazen as the play continues.* **LORIN** *exits and* **LD** *moves to the side to observe the opening of the next sequence.*

A lively choreographed sequence, with appropriate music, showing children (the ensemble) playing outdoor (street) games through the ages, including hopscotch, football, cricket, skateboarding, "paper, scissor, rock", and others. **BRECK**, *dressed with a generous splash of blue (other performers should avoid blue) is playing too.*

LD *enters with a box and issues game controllers to characters as he passes them. They become static as though hypnotised by the screen in front of them. Finally, he approaches* **BRECK**, *who takes a controller.* **LD** *dismisses the others.* **BRECK** *and* **LD** *remain alone on stage. Cold white light flashes as* **LD** *makes to strike* **BRECK**. *Blackout.*

When the lights return **LD** *is alone centre-stage with* **BRECK**'s *family standing away from him.* **BRECK** *is no longer there.*

LD *dials 999 and waits to speak as* **BARRY** *delivers his opening monologue.*

BARRY: If you want to get a feeling of the cold character of Lewis Daynes, hear his 999 call. I mean… come on! That tells you how controlling and calculating he was. My Breck stood no chance, absolutely no chance!

OPERATOR: Essex police and emergency.

LD: Hi there, I need police and a forensic team to my address please.

OPERATOR: What's happened?

LD: My friend and I got into an altercation and I'm the only one who came out alive.

OPERATOR: Are you telling me you've killed somebody?

LD: Yes. I am.

OPERATOR: Right, and who am I speaking to?

LD: My name is Lewis Daynes. I'm eighteen years old.

OPERATOR: ... and what's actually happened?

LD: My friend came to stay the night with me yesterday, feeling down, suicidal this morning. He was in a mess and I tried to calm him down... he shrugged me off...

OPERATOR: Can you just bear with me a second... don't tell me any more... okay? You're telling me he's definitely dead?

LD: Yes. He's definitely dead.

OPERATOR: Bear with me a second.
Right... okay... Lewis... I need you to tell me... what's his name?

LD: Breck. B. R. E. C. K.

OPERATOR: Sorry.... B. R. E...

LD: C. K.

OPERATOR: Breck?

LD: Yes. Breck Bednar.

In silence, **LORIN** *places the stone she is holding with the others.*

LORIN: When I heard this, I was shocked. For one that he admitted that he had killed my Breck. I couldn't believe how he couldn't be sorry or upset.

LD: I can explain this to the officers when they...

OPERATOR: I will... I need to take these initial details.

LD: I know. I know.

OPERATOR: Did you have an argument with him last night?

LD: No, we were fine.

OPERATOR: You were fighting?

LD: No we were fine. Fine. I've known him for a long time. He came to stay with me. I was in Thailand, on holiday, he was in Spain...

OPERATOR: Right, okay, okay, okay. And what's happened this morning?

LD: I woke up and he was in a mess. I put my arms around him, he just shrugged me off, he was going on about how he didn't want to go home and…

OPERATOR: Right okay. *(Typing furiously.)*
Okay. Go on.

LD: In my room adjacent to my bed I have a chest of drawers and a penknife on there, folded. He picked it up, opened it and then lost control… I

OPERATOR: Slow down… let me just… let me take those details… he grabbed the penknife in order to harm himself?

LD: No, in order to harm me. He opened it, then lost control.

OPERATOR: Right… in order to harm the informant. Yeah…

LD: I, in self defence, put my left arm up to block him from stabbing me effectively. We struggled… I got the knife… *(deep breath)* oh, okay… and… can you… can you not interrupt me on this part…

OPERATOR: Okay.

LD: This is being recorded anyway isn't it?

OPERATOR: Yeh.

LD: *(Deep breath.)* I grabbed a knife and I stabbed him once in the back of the neck, I believe somewhere near the brainstem. He turned around to try to carry it on and I… I think I stumbled on my chest of drawers. I fell over. I got up… backed away and… and… I don't remember exactly what happened, but the fight ended with me cutting his throat. I believe I turned around and slashed his throat.

OPERATOR: Right. Okay.

LD: He fell face first on my bed… his throat was properly cut.

OPERATOR: Lewis, are you still in the room where Breck is?

LD: No.

OPERATOR: Where are you?

LD: Look… this… don't… okay don't interrupt me… I didn't know what to do. I felt like taking my own life, I… I don't remember what happened after that… look, are the… are the police on their way? I can hear something.

OPERATOR: They are. They are.

LD: I'm going to go and do my part. Thank you for your help. *(Hangs up.)*

Section 2

LIFE AFTER MURDER

CHLOE: I don't remember going back to school after Breck was murdered. He was, like, in Year 10 and we were in Year 7.

CARLY: I do. We sat in a big assembly and they said:

Don't talk to the triplets about this.

They probably thought we just needed to get on with things, but it was like they were completely disregarding it. I wanted to grieve and remember him. I wanted my routine to be messed up, but it was all just… no one was allowed to talk about it.

CHLOE: I don't remember that at all.

SEBASTIAN: No.

LORIN: At home, all it was, was police and miserable investigations. It was a horrible, horrible time, so we decided it would be best for them to go back to school rather than sit around at home with a bunch of crying adults.

CARLY: I remember, they gave us these little cards to basically, get us out of our lessons. It felt like we were making a fuss, having to get up and disrupt everyone and then go and sit in that room where… it was the same room where all the badly behaved kids went. It was just not a comfortable atmosphere. I was… I'd rather be in lessons.

CHLOE: I remember all my friends asking me… 'cos you could bring a friend with you…

FRIENDS: Can I come with you?

CHLOE: And it was like… I don't want to use Breck's death to my advantage. It seemed wrong.

CARLY: When it first happened, I just turned into this like, really sad girl moping around. Then I got into a really dark place, but in Year 8 I was like… I just made a decision to be happy. I'm not forgetting my brother, I'm choosing to move on. It's no fun being the girl whose brother died. That's not what I want my identity to be.

CHLOE: She gets annoyed because, as one of the triplets, she didn't have her own identity...

CARLY: I used to be... but now it's fun having your own little clique...

LORIN: We decided to move to the coast. I did it because...

CHLOE: Mum wanted a fresh start. That's why, wasn't it?

LORIN: When we came on holiday here, we had so much fun... and, I had PTSD (Post Traumatic Stress Disorder), so it felt like a place to get better.

CARLY: I didn't feel we'd ever fully recover in that house where he'd been right before he died... too many memories attached to the town and the people.

SEBASTIAN: I didn't mind living there.

LORIN: I couldn't leave that house without... it was weird having everyone look at you, knowing what you looked like when you cried here and when you cried there... it was too hard.

CARLY: Mum couldn't just go out into our local area. It was different on holiday. She could have fun... relax. It was refreshing to see that change.

LORIN: I was worried about uprooting you from your school but...

SEBASTIAN: I didn't like the idea of the move at all. I do now... fresh slate. It became a way for me to meet new people.

LORIN: I didn't tell their schools about Breck, because I wanted them to start fresh. They were already "the triplets", they had American parents and then this.

SEBASTIAN: I was glad no one was told. It was a new hostile environment, and I didn't want to kindle something for them to use against me.

CARLY: A really awkward question is: "Do you have any other siblings." I say: Yeh. We have a brother but he's... he's not here any more.

CHLOE: It's awkward in, like, French 'cos they ask how many siblings do you have? Do I say three or two? I say three. Two brothers and a sister.

CARLY: It's kind of like a trigger.

CHLOE: Can I just say how I felt about the move? I didn't want to leave my friends and was quite mad because we didn't know anything about it. I pretty much came home and there was a For Sale sign outside like. *(To* **LORIN**.*)* You didn't tell us, like, anything.

I was upset 'cos I didn't want to move away from my dad *(***BARRY** *enters.* **CHLOE** *looks over to him)* 'cos he'd literally just moved closer to make it easier to see him. I remember checking the train times. It was three hours away!

Being Breck's sister was never an issue for me. I didn't think it was "a thing" living in the same house. I wanted to keep those memories. It was, like, a sentimental house to me so I wanted to stay. I didn't want to move to the coast and I certainly didn't want to go to an all girls school! I was **not** happy about that!

CARLY: She's rolling in boys now and she loves it!

(They all laugh.)

BARRY: The grooming aspect of Breck's murder has been covered on a large scale by the Foundation. Another social concern is to understand how Lewis Daynes got to that point. We need to look at mental health. One of the things I struggled with for a long time was my feeling of compassion for Lewis Daynes, because, I mean, how fucked up was his life, that he got to that point? His mum abandoned him; his father abandoned him; his foster-mother abandoned him. Something bad was bound to happen. He was doomed.

The internet is a tool people access and another way they rise to the top of the froth, as it were. They need help before they do something dreadful like this.

As a father, I feel bad for the kid, you know, because he's in that situation. His life was ruined, Breck's life was ruined, our lives are ruined and, I don't know how you write a play, but I'd like to ask you to have this in the back of your mind, and if there can be some undertone, something about that, I think that'd be powerful.

Section 3

EARLY DAYS

LORIN: Breck's dad and I met in Breckenridge, Colorado. We got engaged three months later and always said we'd name our firstborn Breckenridge. We shortened it to Breck.

BARRY: Breck was a perfect little kid, I mean, your first child makes you a parent. I guess that's a key thing. He got good grades at school He wasn't top of his class which I encouraged 'cos he had other interests, very well rounded.

LORIN: At the time this story begins, I was living in Caterham, Surrey with Breck and the triplets, a perfect place to bring children up away from the gun crime in America.

CARLY: He was so imaginative, coming up with fun games. We'd just never tire spending time with him.

LORIN: Barry and I had been divorced for some time. The triplets and Breck visited him every other weekend.

CHLOE: Breck was, like, the man of the house. He knew how to fix anything.

CARLY: He'd play on the computer and we'd all pull up chairs, gather round and watch the screen.

LORIN: He put all his birthday and Christmas money into his computer and gaming kit.

CARLY: We'd just sit there for hours. I don't know why but it was so much fun for us!

CHLOE: He used to take his computer apart and it'd always be all over the floor.

BARRY: He was always into computer games. I really liked it when he got into his flying simulators and we got him a joystick and all that. He wanted to be a pilot so that's why he got into the air cadets. Then he started getting war games. He'd asked me a couple of times:

BRECK: Can I get the 18+?

BARRY: He was kind of young and I said no, but eventually I kind of let him get one, erm… erm, but he didn't abuse the privilege. I have a flat so, we were kind of, on top of each other so, we'd often go out and do things…

LORIN: He went to Caterham private school in year 7, and loved it. Unfortunately, after two years, his dad said he wasn't able to pay, but Breck had been on the wait list for a state school with a good reputation and, at that exact time, they called and said:

SCHOOL: We have a place for Breck.

LORIN: Meant to be? *(Regretfully.)* Hmmm… so, Breck switched schools. I didn't want to uproot him and felt extreme guilt. He didn't complain, but he didn't find his niche 'cos he didn't have anything to hook onto, until one day he came home really happy…

BRECK: Mom guess what? I ran into some guys I used to play with back in primary school…

LORIN: He listed the names.

BRECK: They've invited me to their gaming group.

LORIN: I thought "Great!"

He'd come home from school, rush through his chores, do his homework and meet the guys online, laughing and joking. I knew these boys, their parents, where they lived, so I considered it a "safe place".

BARRY: "Steam" is a platform to buy and play games on. He had to ask me to put it on my computer at my flat.

LORIN: He had three screens. *(Indicates each position as she describes.)* The gamers were here *(***OLLIE** *&* **MATT** *wave)* kind of a list, and then in the centre was the actual game, and then on the right was sort of, erm… texts, messages and things popping up.

BARRY: He was starting to chat with these friends from the TeamSpeak server.

MATT: TeamSpeak's like Skype without the video. Everyone has a microphone and headset and you can talk to each other, so all we had to do was go online, game and talk to each other rather than going out.

BARRY: ... and they were just kids and, you know, I mean, I worked on the trading floor where there was non-stop banter and I was just like, "Yeh that's great! That's the way we talk!" He just seemed to be happy... but then his mother brought up that there was this guy... LD...

LD, LORIN & BARRY: Lewis Daynes.

Section 4

SUSPICIONS

LORIN: I guess it was January, just over a year before Breck was murdered, when I... I was in the kitchen and I heard this voice.

LD: Impressive, Breck!

LORIN 2: Confident, controlling, taking charge of conversations rather than just hanging out like the others.

LD: Not many guys of your age can do that?

LORIN: I went to Breck's room...

LORIN 2: ... the door was always, you know, open, and I said:

LORIN: Breck? Who were you speaking to then? It sounded like a man.

BRECK: It's Lewis. He runs the TeamSpeak server.

LORIN: He pointed to a picture of a... like a model from a cologne advert.

LORIN 2: Straight away I thought...

LORIN 3: ... he could be some creepy guy sitting in his underpants, pretending to be a teenager.

LORIN 2: He looked too coiffed, too perfect...

LORIN: *(To* **BRECK***.)* How old is he, this Lewis?

BRECK: The guys have been gaming with him for years. Eighteen. He lives in New York.

LORIN: New York???

LORIN 1-3: I kind of left the room thinking, I don't buy this.

OLLIE: Lewis ran a TeamSpeak server and er, seemed like a... like a nice lad...

MATT: ... welcoming, keen to build up a community of like-minded people. We were all in awe of his technical ability.

OLLIE: He'd pop on to say...

OLLIE/LD: Hello. How you all doing?

MATT: He even bought us a game or two!

OLLIE: After a while, I asked him if he'd set up a TeamSpeak server for my friends. He did and gradually it started expanding. By the time Breck joined us, we were a quite large group, around 15, most of us living in Caterham.

MATT: I first spoke to Breck, playing Minecraft.

OLLIE: Soon we met up...

MATT: ... and just sort of hung out for a bit. Breck was into Battlefield 3 and, 'cos he spent more time playing than us, he was much, much better despite me being the year above him at school.

OLLIE: I knew Breck was in the Redhill Air Cadets and he urged me to join, so I spoke to him there as well.

LORIN: A predator will infiltrate a group and kids won't realise; nobody actually knows him.

OLLIE: Lewis told me he was a defence contractor, with work and important meetings in Washington.

LD: I'm testing security for the US government. By day I run a computer technology company.

MATT: What, like hacking?

LD: Can't say any more... you can't tell who may be picking up these communications online.

OLLIE: He claimed he made loads of money.

MATT: Lewis gradually became Breck's mentor, and he soaked up information for a career he wanted desperately.

LORIN: Some of the other gamers were sixteen or seventeen, so they wouldn't've felt he was an "older person".

LORIN 2: Breck never considered LD 'a stranger'. He trusted his friends.

LORIN: I should say here that I don't like to say his name, so I use his initials. (In my head it means "limp dick".)

LORIN 3: I google-stalked Lewis Daynes and found profiles saying his name and age…

LORIN 2: … but no photos.

LORIN 3: He did let them see one… just once…

LORIN 2: … and that was enough for them to let their guard down,

BRECK, OLLIE, MATT & LORIN 1-3: He's not some sort of creepy old guy. He's practically the same age as us!

LORIN 2: He wasn't the stereotypical version of what a predator is.

LORIN: A lot are the same age. It doesn't mean they're safe.

LORIN 3: I thought, I need to get to know him, so I'd kind of, go in and say:

LORIN 1: Hey guys what's going on?

LORIN 3: And he'd chat to me.

LD: Hey Mom what's going on? What are you doing?

LORIN 3: And I'd say:

LORIN 1: Just back from a dance class. I'm exhausted!

LORIN 2: Dancing ladies would pop-up on the screen… that quick!

LORIN 3: If I'd been to the cinema, straightaway a trailer would pop up.

LORIN 1: Things happened so fast, I could see why Breck liked to interact this way.

LORIN 2: Funny things.

LORIN 1-3: … like a virtual clubhouse.

LORIN: That's what Barry used to say.

LORIN 3: The other teenage boys would get, like, stiff and quiet when I started to talk but LD would show off, so I tried to get information from him:

LORIN: *(To LD.)* So, have you been out this weekend?

LD: No. Me and my girlfriend just split up and I was working.… really busy.

LORIN 3: And I'm thinking…

LORIN 2: Yeh, you're so busy…

LORIN 1-3: … yet you're always online with my son.

LORIN 3: So many teenagers over-share online…

LORIN 2: … yet this LD guy was being so super-careful and hiding his identity.

MATT: Unbeknown to Breck's mum, Ollie, and some of us older ones, were beginning to suspect Lewis wasn't all that he made himself out to be.

OLLIE: I mean, I was going to an air show. I went every year with friends, some who I'd met over the internet. It was safe, in a public place and I asked Lewis if he wanted to join us. He declined:

LD: I'm busy with work…

MATT: Every time we planned something, he'd be like:

LD: I can't do that, I'm in America.

OLLIE: … or

LD: I'm seeing family.

OLLIE: Then, when we'd Skype, he didn't have a webcam. That was odd.

LD: It's not working.

OLLIE: … or…

LD: I've been so busy I haven't ordered a new one yet.

OLLIE: It just kept happening.

MATT: We asked him to send photos and he sent fake ones.

OLLIE: One time he told me he had to get a flight back from Washington on a large military aircraft and they had a problem over the Atlantic, so they made an emergency landing on an aircraft carrier. Now, me being a massive aviation geek, I just knew that was physically impossible. It was a flat out lie.
"We'd've heard about it on the news though, Lewis!"

LD: No, no no! 'cos it'd be such an embarrassment to the government!

OLLIE: I just thought, that's bollocks! Maybe he's doing it so we like him more or to seem… like a cooler person.
Then I started to think… mmm, maybe you're not a defence contractor either?

LORIN 2: He never had New York information or stories and was never in US time zones:

LORIN: Jeez, he's never working!

BRECK & LORIN 3: He works his own hours, on contract.

LORIN: What does he do exactly?

BRECK: You won't believe me.

LORIN 1-3: Try me.

BRECK: He works for the US government, in the FBI. Online surveillance, that kind of thing. That's why he can't show his identity. He works when he wants to work and he…

LORIN: How old did you say he was?

BRECK: Nearly 18.

LORIN: How did he get that level of job in the States when he's so young… and he's English?

BRECK: He's that good on the computer, Mom!

LORIN: It just doesn't seem…

BRECK: I wish I could have his life. Travelling round the world working on a computer all the time and earning loads of money. I'd love it! He's amazing! You should be grateful we found each other!

OLLIE: I've met and made good friends through gaming and then, from that, I'd Skype them, see their faces, hear their parents, their brothers and sisters, so I know they're just another teenager sitting at their computer, chilling, playing games. So, I'm not saying everyone out there is a paedophile but I have also been in the situation where someone isn't who they say they are, and that led to the death of my friend. I knew both the murderer and the, oh what's the word… the victim. I considered myself a friend of the murderer as well. That's a scary thought.

I was the first person in the gang to meet Lewis, a year before I met Breck, so I kind of feel responsible for what happened because, you know, I introduced them.

LORIN: The boys seemed to be impressed and liked being involved with someone who was, you know, so high up in his career path, yet was playing games and giving his attention to them. I may not have been tech savvy, yet relationship-wise, I knew this didn't look like a "healthy relationship".

BRECK: Mom, you're paranoid! Maybe I should just stop telling you stuff.

LORIN: I... I felt I had to be careful.

LORIN 2: I didn't want Breck to stop talking.

LORIN 3: Why didn't he think it sounded suspicious?

LORIN: Their relationship was soon starting to hurt my ability to reach Breck. He was an A* student but also had common sense. In the summer of Year 9, his dad got him a job in the city doing data processing. At the end, I remember him saying, and this is a key point...

BRECK: I'm never doing that for my job, Mom. I want something exciting!

LORIN: Coming into the Autumn term Breck's personality changed. I couldn't tell if it was teenage stroppiness...

LORIN 1-3: ... or something more.

LORIN 2: He questioned everything...

LORIN 3: ... not just questioned, but told me what LD thought...

LORIN: Don't forget to do your chores.

LORIN 2: ... or...

LORIN: Rubbish and recycling night.

LORIN 3: He'd literally just stand up tall and go...

LORIN 4 & BRECK: Huh! Lewis says you shouldn't force me to do chores. It's the triplets who make the mess!

LORIN 2: And I'd be like...

LORIN 4: We all pitch in. It's just normal with a family.

LORIN 2: Then on Sunday...

LORIN: Breck, get ready for church, we're leaving in half an hour.

BRECK & LORIN 4: I don't believe in God anymore! Lewis says you shouldn't force me to go to church.

LORIN 2: And I'd say…

LORIN 5: What has this has got to do with Lewis?

CHLOE: I was like "yes!" Listen to him, Mum!

CARLY: No, I didn't want to go to church either!

SEBASTIAN: I just wished I had the nerve to challenge Mum like he did.

CHLOE: I've been raising my voice about church ever since!

SEBASTIAN: Who wants to be forced to go to church? In the end it turned out to be rather fun!

CARLY: His attitude to Mum changed.

CHLOE: It wasn't like one day he was, like, super nice and the next he was horrible.

CARLY: He'd been respectful to her before and like… he just never wanted to hang out as a family anymore which was kind of… it just felt rude.

CHLOE: I assumed it was him having hormones and that's why he was shutting us out more.

CARLY: I didn't say anything 'cos I was frightened it would cause, erm… arguments. I got scared when they were fighting.

SEBASTIAN: It was none of my business, nor did I care to make it mine.

BRECK: Lewis says I don't drink, I don't smoke, I don't do drugs, so why won't you let me spend time on the computer.

LORIN 4 & BRECK: Lewis says I don't cause problems so why are you punishing me?

LORIN 5 & BRECK: Lewis says I'm clever and could go places. Why're you holding me back?

LORIN 2-5 & BRECK: Lewis says I can get a Microsoft apprenticeship through him.

LORIN: It would have been different if LD was someone in the community where I thought:

LORIN 1-5: He's a clever, strong person.

LD: Breck, you don't have to do what these judgemental parents say.

BRECK: I wish I didn't.

LD: You don't. You're a clever dude. You have so much potential.

BRECK: Really?

LD: Yeh. Your brain is wired the right way. It's rare. You're special.

BRECK: That's so… especially coming from you… it's…

LD: I mean it. You shouldn't waste your time on all these idiots. You can bypass all that. Look at your mum. She's just a teaching assistant in a school. She can't even pay her bills.

LORIN: I spoke to Breck's school. They had no concerns.

TEACHER: I wish all kids were like him.

LD: We've got a special bond, Breck. That's all that matters.

LORIN: I struggled to get Breck to come when I needed him, or to go to bed… and when I went into his room to talk to him…

LORIN 3: … a wicked witch on a broomstick'd fly across the screen …

LORIN 4: … or Hitler would goose-step across as if I was a…

LD & LORIN 4: … mean, domineering dictator.

LORIN 1-5: They were gaming with this 'adult' who had no rules…

LORIN: … so, in the end, I had to yell, to get him off the computer…

LORIN 1-5: Breck!!! Breck!

LORIN: The police told me LD recorded my yelling and looped it to playback to Breck.

LORIN 4 & 5: Breck! Breck! *(Continues under the next three lines.)*

LORIN: Breck was torn…

LORIN 2: … hearing this through his headphones…

LORIN 3: Do I listen to Mom or…

LORIN 4 & LD: … do I listen to this great amazing guy…

LD: ... who's teaching me so much and offering me the world.

LORIN: He had this, you know, conscious...

LORIN & LD: ... subconscious...

LORIN: ... person...

LORIN & LD: ... literally in his ear...

LORIN: ... telling him...

LD: You don't have to do anything.

LORIN: It was a sort of brainwashing.

CARLY: Lewis was like, a serious thing between Mum and Breck.

CHLOE: I hadn't seen any signs of grooming.

LORIN: Grooming works through shared interests.

CHLOE: It's so hard to recognise.

LORIN 3: If these boys weren't techies they wouldn't've spent time with him.

LORIN 2: He was grooming hundreds of boys all over the country through Roblox, Minecraft, Battlefield, FIFA and Call of Duty... any game that's out there.

LORIN 4: He'd build relationships...

LORIN 5: ... then focus on one boy...

LORIN 4: ... get bored...

LORIN 5: ... then another.

LORIN 1-5: Breck was his final target.

LORIN: Part of me wonders if Breck was a bigger challenge 'cos I was the mom standing in his way.

Section 5

ISOLATED

CHLOE: I remember on the way back from air cadets one day, there was a big argument. I was just in the car, in the back, really scared 'cos *(To* **LORIN**.*)* you were shouting at Breck.

CARLY: We had no idea how big this issue was. I just wanted the fighting to stop. I was like "Come on guys".

SEBASTIAN: Breck grew reclusive and angsty. He was upset, and we could see it.

LORIN: At Thanksgiving that year, we were invited to a friend's house. Breck didn't want to be there and openly showed his disdain... the body language, the face, and under his breath, saying something like...

LORIN 2-5: I don't want to be here. This is shit...

LORIN: That's not the exact quote but it's what his face was saying. That was the first time I noticed him being openly rude to other people, not just me.

LORIN 3: We didn't know then, but some older friends had seen through LD and had actually set up a new server...

OLLIE: Lewis had become controlling. He banned my friend, we'll call him 'Will'. *(***WILL** *enters.)*

MATT: Will realised Lewis was lying about things, like being part of the military and wanting to go to Syria to fight there, he thought...

WILL: Hang on, if he's, like, 18, you can only join the military when you're 16, how's he done that as well as college?

MATT: And, er, it confirmed my suspicions. Once we started to suss him out, he sort of lost it a little bit.

WILL: He's power crazy! Watch when I speak up against him.

LD: *(Menacing)* You respect me! I'm older than you. You don't argue with me.

MATT: And Will would, like, continue and Lewis would mute him. Then ban him.

MATT & OLLIE: What happened there?

LD: I banned him.

OLLIE: What the hell?

MATT: I remember feeling a bit sort of scared about speaking up, so Lewis would be almost, like, overly nice to me. Breck had an interest in, like, programming and Lewis would give him private coding lessons on a separate channel. If I tried to join he'd be like:

LD: I'm talking to Breck.

MATT: I felt left out and thought... why can't he teach us both at the same time?

OLLIE: Then he revoked my powers on the server he'd built for me and gave them to Breck and others, so, I was like: I'll build my own server.

MATT: Looking back, erm, I think Lewis skipped between people. There was someone called Josh who Lewis was close to, and I remember him saying he felt betrayed 'cos Josh answered back. When he banned me Will said:

WILL: Me and Ollie have got our own server, do you want to come?

MATT: I remember sending a message to Breck saying:
"If you want to talk to us we're here."
... and he said:

BRECK: Thanks. Maybe at some point.

MATT: He never stayed long... Lewis always came to listen.

One day, I remember going onto our TeamSpeak... and all our channels had been deleted. The only ones left were named...

LD: You've been hacked!

MATT: Or

LD: This is really easy. You should have better security on your server.

MATT: Even if we'd wanted to ban him, we wouldn't 'cos he'd just say:

LD: *(Laughing.)* Breck, look, they've banned me from their little server.

MATT: It was a tricky position.

OLLIE: The majority moved with me. Breck was very much invited but stayed with Lewis 'cos he was being favoured, getting free games and software.

LORIN 5: My love was beginning to be no match for the dark power LD exerted over him.

BRECK: Mom, I'm not going to air cadets any more.

LORIN: Why, honey? What do you mean?

BRECK: There's guys there I'm not getting on with. One's been saying stuff….

LORIN 2: This was a best friend from primary school!

LORIN 3: He stopped going.

LORIN 4: LD had stirred it all up.

LORIN 5: He was gradually…

LORIN 1-5: … isolating Breck.

BRECK: Lewis made $2 million in bitcoin trades.

LORIN: What?

BRECK: And he gave it to the Syrian rebels.

LORIN 5: I was like… maybe they're being radicalised.

BRECK: *(To* **LD***.)* Syrian rebels? Are you sure this is?

LD: It's not the first time. I supported the Egyptian rebels too… wanted to fight for them but… my money can assist more…

LORIN: $2 million? That's a lot of money.

BRECK: He's such a good trader.

LORIN: And he gave it away?

BRECK: That's the kind of good, helpful guy he is!

LORIN: He's does this and works for the US government?

BARRY: Breck, come on! This is bullshit!

BRECK: He's a good guy, Dad!

BARRY: The US government doesn't just give out passports for this stuff. The guy is full of shit!

LD: It's abuse, Breck.

BRECK: I don't think...

LD: It's what it amounts to, but just give me a minute. I'll concoct an email. I'll write it, but it will come from you. You can't let them get away with this. It'll shut them up I promise.

BRECK: It'll make things worse.

LD: Trust me. They'll roll over.

BRECK: I'm not sure.

LD: You got to be strong, Breck, otherwise they'll... it'll carry on... and it'll stop you from reaching your potential. Our bond, Breck. Remember our bond!

BARRY: Then Lorin gets this email...

LORIN: ... a manifesto...

BARRY: ... from Breck.

LORIN: ... two pages, with all the things that he should and shouldn't have to do, and at the end...

"... if you don't listen to me, my good friend Lewis will get me a Microsoft Apprenticeship when I'm 16 and I'll be outta here!"

*(To **BARRY**.)* Barry, I can't have this threatening tone... then... when I printed it to show you, it kind of backed up to a page saying:

Written by Lewis Daynes. Edited by Lewis Daynes.

BARRY: What?

LORIN: I think I need to call the police.

Silence.

BARRY: Right. Yeh.

Silence.

And she did. We were a pretty... a very concerted front.

LORIN: It was the biggest... most important thing I could do.

LORIN 2: A friend came over to support me 'cos I was so nervous.

POLICE: *(On phone.)* How can I help?

LORIN: I have a 14-year-old son and I feel like he is being groomed online by the guy who runs the server they play games on.

LORIN 3: I handed over his name, and an alias Breck had given me...

LORIN: I don't know if either are real.

POLICE: We'll make intelligence checks to see if he is known to anyone else through the Police National Computer. You must take away your son's computer and then ensure he uses a different server.

LORIN: I came off the phone, relieved.

LORIN 2: The records would be checked.

LORIN 4: I went to his room and took away his technology.

LORIN 5: I hid it in my lodger's room.

ALL LORINS: Breck had no access.

CARLY: I remember thinking that was harsh. I didn't realise it was for a 'dangerous' reason. I thought it was a punishment for not doing his chores. Breck was upset, and I remember going to Mum saying: Please give it back. It's his life!

LORIN 5: Had the police popped LD's name into their computer they'd've discovered Daynes was known for prior allegations of sexual activity against a 15-year-old boy in 2011, and for indecent images on his computer.

LORIN 1-5: Unfortunately, they didn't.

BARRY: Lorin also told the police where Daynes lived. Essex and potentially Greenwich, so another data point was ignored.

LORIN: In my pretend dreams the police would come to the door, speak to Breck and say...

LORIN 1-5: "Your mom was right."

LORIN: … and Breck would finally believe me. That should have happened. Breck would still be here and I'd be a productive human being instead of what I feel I am now…. a shell of a person.

Section 6

DRIVEN UNDERGROUND

BARRY: We decided to have, for the want of a better word, an "intervention".

Breck's closest friend from school was a guy named Joe. His parents were about as conservative as you'd ever find.

LORIN: They were alarmed by what I was saying, but when I asked if we could get more parents they…

LORIN 3: … they didn't want…erm… you know…

LORIN 2 & 3: … the stigma of having the community knowing their business.

LORIN: So, just our two families met.

LORIN 5: We found out later, LD had instructed Breck to secretly record the meeting, to know what we'd said.

LORIN 1-5: This is from that recording.

LORIN: *Breck doesn't like the word 'influence' but I do believe he was influencing Breck and maybe others. It might sound like paranoia, but I felt like it was even grooming behaviour.*

BRECK: *It's a harsh term to say grooming.*

BARRY: *That's why we're concerned, Breck. And the stuff we hear scares us to death! It just sounds incredibly odd… working for a major consulting firm in the US when you're sixteen years old!*

BRECK: *Seventeen.*

BARRY: *It's fanciful, and what that does is raise a flag to us that something else is wrong here.*

BRECK: *It's gonna sound weird but there is no one else who sort of, I share interests with.*

BARRY: It was not friendly I can tell you that. I remember saying: *We love you and we want to make sure you're safe. You need to stop having contact with him unless he agrees to meet us and allay our fears.*

BRECK: *He won't ever meet up with any of you!*

26

BARRY: At the end of the recording, we hear Breck say:

BRECK: *I don't really see the value of worrying about this anymore. I will do what you want. I will sort of, distance myself from Lewis. It's not worth, er, getting too involved with someone you don't know in real life, I guess.*

BARRY: *So, going forward, I think we've determined that Lewis is kind of out of the picture today.*

BRECK: *Yeh.*

LORIN 1-5: We thought we'd fixed it.

BARRY: I remember on the way home I said to Lorin:

That was too easy.

LORIN 5: Unbelievably *(sigh)*, LD had secretly couriered Breck a brand-new iPhone. I didn't see it arrive. The whole thing went underground.

LD: Breck… your parents are treating me like some kind of criminal creep. No one does that, dude.

LORIN 1-5: … and underground is exactly where a predator wants it.

BRECK: Don't worry. No one will break our bond.

LORIN 4: I nearly… his… one day his pocket suddenly lit up… and I went…

LORIN: Breck? What's that in your pocket?

LORIN 2: And he's like…

LORIN & BRECK: What?

LORIN 2: And before I could form a coherent sentence he said:

LORIN & BRECK: Oh… that's Tom's. He let me borrow it.

LORIN 2: And I went…

LORIN: How does Tom have a spare phone?

BRECK: Well, because he just got a new one so it's spare.

LORIN: Give it to me.

LORIN 3: So, I take it. I…

LORIN 2: … I put it in an envelope.

LORIN 5: This is, you know, if you could really show how stupid a person is.

LORIN 4: I seal the envelope…

LORIN 5: … label it "Tom"…

LORIN 1-5: *(In amazement.)* … and hand it back to Breck!

LORIN: You take this back to Tom and don't borrow it again.

LORIN 5: That's how much I trusted him.

LORIN: Breck was, I thought, absolutely back to normal. He did his chores, didn't complain, was more respectful of how the house ran and it seemed like he'd understood why we'd taken everything away. I never heard that deep and distinctive voice again.

We had a great Christmas…

BARRY: I remember a picture of Breck, reading a book, not on a computer, not on his phone. It was like he was sort of reborn, so I said to Lorin:

I think he deserves to get his stuff back. He's learnt his lesson… so, we… we did, but we told him:

BARRY & LORIN: You mustn't speak to LD unless he agrees to meet us.

LORIN: On Boxing Day Breck got his technology back. He wasn't like whoo hoo! I tried to empower him and sort of said: Breck… you can make this new server better than LD's!

LORIN 5: Joe's parents found him back online with LD in January so kicked him off again. They didn't pick up the phone, or send a text to…

BARRY: … a breach of trust and responsibility in our opinion…

LORIN 5: … 'cos I could've spoken to Breck…

LORIN 3: He might have admitted it or not…

LORIN 2: … but it would have prompted me to call the police again.

LORIN: I didn't, 'cos there was nothing new to report.

LORIN 5: In the investigation, the police asked me why there were three routers in Breck's bedroom.

LORIN 4: Even if I'd turned our WiFi off, at night, he and Lewis could still talk.

MATT: After Christmas we didn't have much chance to speak to Breck as he didn't come onto our server. We weren't at the same school, but Joe was and I remember him saying...

JOE: Breck's blanking me.

MATT: ... and we thought

"Okay. He obviously doesn't want to talk to us."

... and we started to get annoyed with him, you know, choosing someone he hadn't met before, over us.

OLLIE: On Lewis's server we couldn't just go in and talk to Breck, but we did convince him onto our server where about five of us tried to tell him Lewis wasn't who he said he was. Breck was dismissive.

BRECK: He's alright. He's chilled. He's really nice to me. Look, Lewis is messaging me, he wants me to go back.

MATT: We had no more conversation 'cos he didn't come onto our server again.

Section 7

PLANS

OLLIE: The vibe I got was, Lewis was dodgy... I could never pinpoint it. Murder was never in my mind but... yeah, possibly paedophilia.

LORIN: ... as a BBC documentary later revealed, LD was sending messages to Breck... including this:

LD: Breck, I'm sick. I'm dying. I want you to take over a contract worth between £55 and £135 million. I have a buyer, I can't say who, needless to say it's a government entity that would like to operate as a subsidiary of a larger organisation and keep a team of privately employed technicians on hand. I don't think I have to say who that money will go to do I? (You). I need you to read some documents to help you understand what's needed. I'm literally trusting you with my life's work. You have so much to learn and not a lot of time. You're the only person I can trust. You'll be successful beyond your wildest dreams and will have financial security forever. You're walking in the real world now, Breck. I may not make it. If I don't, here's the number of the guy at the US embassy. He'll hook you up with all the relevant information because I'm really... Oh sorry, Breck, I'm weakening now... I'll explain everything when we meet up. A taxi will pick you up on February 16th. Breck... book a pre-paid taxi from your house to this address in Grays for that morning and we can go over everything.

LORIN: When February came, I thought, I've got everything going for me.

LD: I'll transfer the money to your account.

LORIN: Breck's back to normal, the triplets are happy in school, everything was going well at home, I was interviewing for a new job. I felt good. I looked good.

LD: Yes. Sunday 16th February. Say 7am before everyone's up.

LORIN: Breck was off on a Spanish exchange trip just before the February half term...

LD: Do it before you leave.

LORIN: ... a perfect opportunity to spend time, face-to-face with his friends, so he wouldn't be tempted to fall back into the trap of thinking LD was his real friend.

LD: Amazon order: Duct Ultimate Cloth Tape, Silver – 50 mm x 25mm.

LORIN: He was getting ready for this trip the night before...

LD: Durex. Topsafe Condoms – 12 pack

LORIN 2: ... and he was **really** preoccupied with his computer...

LD: Multitool pocket knife.

LORIN 3: ... like it was completely taken apart and I was thinking...

LD: Buy now with one click.

LORIN 2: How come this is happening now?

LORIN: He was working, chatting with me but, absentmindedly.

LD: I'm in Dubai on a layover, departing at 3pm. I don't know if I'll get the chance to speak to you before you go. Have a go at flashing the SSDs. Hopefully that will solve issues. If not don't stress about it, we'll sort it on the 16th.

LORIN: Do you want me to help you pack? I knew I wasn't going to see him *(voice breaks)* for a while...

LORIN 3: I didn't know how long "for a while" was going to be... .

LORIN: ... I thought it was going to be two weeks, because I was also away in Spain but not near him.

LD: Enjoy your time in Spain even though it's a school trip, just have fun and relax. Don't feel awkward around people. I'll try to email if I can during my trip, but if I don't it's because I can't get a connection, so I'll speak to you on the 15th, when we both get home. Have a good one and enjoy yourself. Lewis.

LORIN: Mine wasn't a "whoohoo" holiday, I didn't have money for real holidays. I was literally in Madrid, sitting in a lounge speaking English to two Spanish students, giving them practice talking but, I was away for my birthday...

LD & LORIN 1-5: ... the 17th February.

LORIN: I stuck in a little bon voyage card, some chocolates, a gift and a thank you card for the host family.

Next morning, he was getting a lift with a friend to the airport. I took a picture of him. He didn't want to look at the camera because he didn't like being photographed. He walked out of the door so tall and handsome. I was so proud but never expected that to be the last time I'd see him alive.

We hugged and kissed and said we loved each other. It was a nice moment and off he went.

I watched as they drove away just, you know, he was like... he was my right-hand man 'cos, you know, boyfriends come and go but, Breck was always a help to me. I'd tried for years to have him and he was the one that made me a mom when I wanted to be a mom so much. I adored that kid... adored him! He texted me after he was there for a day or so and said:

BRECK: Mom, it's amazing. The host family are sweet. The food is really good.

LORIN: He posted a picture on Facebook, leaning his head in with a girl from the trip. He'd never done that before, ever and, like, I thought: Yes! He's out there experiencing the world. I couldn't think of a better scenario.

LORIN 5: What I didn't know was LD was sending him non-stop messages...

LD: Hey Breck, I tried to call you.

LORIN 5: ... smothering him

LD: Did you get my message?

LORIN 5: ... and isolating him...

LD: Breck, I know you're busy. I need to speak to you. Call me!

LD & LORIN 5: I sent you emails. You need to look at them. Urgently!

LORIN 4: The police said LD was inundating him with information, so much it would have taken an adult weeks to read...

LORIN: I wasn't due back until the week after Breck returned, so he was going back to his dad's.

BARRY: I met him at the airport and when he came through those doors he was all tanned and happy, and looked taller and confident and I was like… ah you know, this is great! There's a huge Tesco right there and he was buying all sorts of stuff to eat that week. He was in great spirits.

LD: If your father asks where you are going on Sunday say you're meeting a friend who is 14, name: Edward Bligh. He invited you over to his dad's house for the day while he's visiting his father who lives in Caterham. Edward Bligh's father also said you might be able to stay the night, depending on their plans.

LD: Also, in casual conversation, just say I've spent too many half terms and holidays indoors, I'm going to spend this one outside with friends. Just some ideas, but it's better than getting caught out if your father asks you to provide details and you slip up.

BARRY: He didn't need to come up with all this shit but that's how in-depth he was.

LORIN: Sadly, Breck was clever enough to just say:

BRECK: Do you mind if I go over to Tom's tomorrow?

BARRY: I'd met Tom; and the one I thought it was, is a good guy

LORIN: There were two Toms in Breck's life. Tom was a safe name.

BARRY: I… I asked him…You want me to drive you there?

BRECK: I'll take care of it.

BARRY: Sure you're alright?

BRECK: Yeah, yeah. All sorted!

BARRY: And I was like:

(High pitched, not wanting to offend voice.) Okay, and I'm happy he's going to see a *(chuckles)* physical friend.

I was staying at Lorin's house which I never normally did, but it was easier 'cos with all the kids, my place was smaller.

I was sleeping on the couch in the, er, living room and I remember the door going and I said:

Breck, you going now?

BRECK: Yeah. See you later.

BARRY: It was early... but I was like, maybe they had plans.

To my lament, I didn't grill him, I was just happy he was going.

We all went out to London.

LORIN: Nobody questioned him 'cos he's not a "vulnerable child". He's a clever, good boy. I mean, a taxi at 7 in the morning? He trusted Breck.

LORIN 4: When he arrives at LD's flat he'd see it wasn't a millionaire's flat but LD, a psychopathic liar, would have said:

LD: This is an undercover work flat. I'll show you my nice place if I get the chance.

BRECK: I thought you'd be more... you know...

LD: The treatment seems to be going well but... well you can never predict.

BARRY: I sent a couple of messages to Breck and he goes...

BRECK: I'm fine. I'm okay.

BARRY: Then he... *(Pulls out a mobile phone.)* I can show you, 'cos I still have them. I don't know to this day whether Breck sent them but...

LD/BRECK: Tom has pizza. Sure you want me to come back?

BARRY: Can you stay all afternoon then?

LD/BRECK: I could even stay overnight. They asked if I could, and I have stuff.

BARRY: Okay, Breck. I need you to come back tomorrow, mid-morning as I have to go to London for lunch and I'd like you to be here. Please be home by 10:30 tomorrow.

Is your zip card here that Chloe can borrow?

LD/BRECK: I have my wallet with me so, no.

BARRY: We'll come and pick it up.

LD/BRECK: I just checked my wallet. It's in my suitcase somewhere from the holiday.

BARRY: I thought you unpacked your suitcase.

LD/BRECK: I did but forgot about that. It's probably in a side pocket. If you can't find it just buy a ticket.

BARRY: It's not in your suitcase. Any other ideas?

LD/BRECK: Why does Chloe keep losing it?

BARRY: No idea.

LD/BRECK: I'm going to install a new operating system on my computer. I built it with a friend for a computing project in school. It's the first real software I've built. I'll call you tonight and if my phone messes up I'll show you tomorrow when I get back. Big smiley face.

BARRY: Okay, be home by 10:30.

LORIN 5: Breck's phone went dead early in the day on the Sunday.

LORIN: Barry didn't think anything of it…

LORIN 2: … a hand-me-down 5 and the battery's run out…

LORIN & BARRY: … they're always dead.

Section 8

MURDER

LORIN: I wake up in Madrid on my birthday…

LORIN 5: … with an impending sense of doom…

LORIN 4: … and I didn't know why.

LORIN 2: Is it because I'm another year older?

LORIN 3: … another grey hair?

LORIN 2: … another wrinkle?

LORIN: That's not impending doom… that's just disappointment.

LORIN 2: Then I thought, is it because the kids aren't here to give me their cards and make your birthday fun?

LORIN 1-5: I couldn't figure out why.

LORIN 5: Then I get this text from his dad:

BARRY: So mad at Breck. He's supposed to be home at 10 to babysit the triplets. He hasn't shown up and his phone's dead! *(Exits.)*

CHLOE: Dad had to go to work, so us three were with some of our old school friends from primary.

CARLY: At a park, a 5 to 10 minute walk from our house… really close…

CHLOE: Everyone kept ignoring me and I was getting upset, so I went home.

CARLY: It was a common thing for her to storm off…

SEBASTIAN & CARLY: … so we let her go.

CHLOE: I was putting my bike away and when I came back to the front of the house there was a police car and they got out and walked up to me:

POLICE OFFICER: Is your brother at home?

CHLOE: And I was just, like:
He's at a sleepover.
And they were, like…

POLICE OFFICER: Where's your parents?

BARRY: I got a call and, at first, I didn't even pick it up 'cos I was in a meeting and it was a mobile number I didn't recognise, then they called again and…

POLICE OFFICER: It's the police. Is this Barry Bednar?

BARRY: Yes.

POLICE OFFICER: Could you come home? We need to talk to you about your daughter, Chloe.

BARRY: What's happened? Is it an emergency?

POLICE OFFICER: You need to come home right away.

BARRY: There had been bullying stuff at the park, but I thought… Chloe's not a tell-tale that'd go to the police, so it must've been something crazy so, ermmm… I got on the train and went home.

CHLOE: Dad called me and said…

BARRY: Did you call them just 'cos your friends were being mean to you?"

CHLOE: No!

BARRY: Why are they there then?

CHLOE: I couldn't give him a solid reason to say, like, why. Then I called Carly and Sebastian and was like:
There's police here, you need to come home.

SEBASTIAN: I was wondering why she'd got the police!

CARLY: I was, like, kind of excited…

SEBASTIAN: I remember going in and… really nice ladies, police officers just… just asking us questions, just, er… keeping us calm I suppose…

CARLY: It felt like we were there for hours and they just kept asking us questions.

SEBASTIAN: … like what are my hobbies… what am I interested in?

CHLOE: I remember one asking: What was our favourite hamburger… so, we sat there naming hamburgers! It was super-uncomfortable.

CARLY: It never crossed my mind it was Breck. He's just not a person you'd worry about. He'd be looking after us if anything.

CHLOE: I remember seeing people in white suits and a lot of police cars pulling up. Then, Sebastian received a text.

SEBASTIAN: It was just a girl, she just said, er… they… something about…

CARLY: It said…

GIRL: Is it true what happened to your brother? If it is, that's very sad…

SEBASTIAN: I was like… what you talking about?

CARLY: I think we… we, like, understood that we weren't supposed to see this… so he just put the phone away and we, like… didn't think we should tell the policewomen. I couldn't wait till Dad got home.

BARRY: I called, er, Lorin and said:
Do you have Tom's number, 'cos Breck still hasn't called me.

LORIN: What do you mean?

BARRY: He's still not back!

LORIN: Something's wrong.

BARRY: Then I realised… this must be about Breck!

LORIN: I'll ring his friends.

MATT: It was half term. I'd been in school catching up and I remember coming home, thinking: That's been a productive day. I went onto the server and I could hear Will talking to someone in the background, and Joe said:

JOE: I need to talk to you about something. It's important but give me a second.

MATT: He sounded really upset.

OLLIE: I popped onto TeamSpeak like I did all the time, and Will rushed into the channel and just said:

WILL: Ollie, Lewis killed Breck.

OLLIE: *(Chuckling, embarrassed.)* What do you mean?

WILL: Stabbed him! Lewis has killed Breck…

OLLIE: I was just trying to work out what he meant and, realising he was genuinely serious I, yeah, I just sort of, just sort of… I was just in shock.

MATT: What do you mean? How did it happen?

WILL: Lewis says it was an accident.

MATT: I remember just throwing off my headphones and running downstairs, and before I even opened my mouth to say a word to my parents, I just burst into tears. They knew something bad had happened 'cos it was just, like… I couldn't get a word out. *(Laughs.)* I started to get my thoughts together. The shock went to sadness and then to, like, just anger.
How? How could he "accidentally" murder someone? I just sort of, left the house and went to Joe's.

JOE: One of the people on Lewis' server came on and said Lewis told them he'd accidentally killed Breck. He said:

LD: Breck came at me with a knife so I tried to disarm him and through the hustle of trying to get the knife off him… I cut his throat.

MATT: How can he "accidentally" do that?

JOE: Then Lewis sent them a photo of Breck's body asking:

LD: I'm showing you this in the hope you can advise me on how I might help poor injured Breck.

MATT: Shit!

LORIN & LORIN 2: His friends kept saying he's with Tom.

LORIN 3: … but then one said…

LORIN 4: Tom's on holiday.

LORIN & LORIN 4: I'm getting more and more worried.

LORIN 1-3: Finally, one said:

ANON GAMER FRIEND: He's unwell. Really unwell.

LORIN 1-3: His voice was quivering.

LORIN: Where is he?

Pause.

Please. Tell me where he is!

ANON GAMER FRIEND: He's with Lewis.

LORIN & LORIN 5: My heart sank. This is the first time 'cos I thought Lewis was completely out of the picture.

BARRY: When I got home I expected… I was like… I wasn't putting two and two together…

CHLOE: All we heard was shouting!

BARRY: The police sat me down and said:

POLICE OFFICER: We have to tell you… Breck is your son, right?

BARRY: Yeah. What's happened.

POLICE OFFICER: He is deceased.

BARRY: That's what he said: "He is deceased".

What? What on earth are you talking about? Did he get hit by a car?

POLICE OFFICER: We believe he's been murdered.

BARRY: And I was like… and so… and so, I had to call Lorin… *(calling on his mobile)*

LORIN: I remember exactly what I said…

BARRY: I don't know how to tell you this, but Lewis has killed Breck.

LORIN: I couldn't stop screaming.

> **LD** *aggressively manhandles* **LORIN,** *with her remaining unaware of her presence and unaware of why her body is reacting in this manner.*

LORINS 1-5: *(Screaming, falling to the floor and saying in no particular order.)*
Oh my God!
Breck's been killed. I'm dead too. I'll never survive this. Take me to the airport.
Get me on a flight!
I tried to warn him.

I did everything I could.
I really thought it was fixed.

LORIN: They called a doctor to sedate me.

At three or four o'clock in the morning a taxi arrived and we went to the airport for the first flight out. I barely got on 'cos I looked like an absolute crazy zombie. The flight person looked at my boyfriend and said:

FLIGHT ATTENDANT: Is she okay to fly?

LORIN: ... like I was some lunatic or something.

All **LORINS** *fall silent.*

CARLY: Dad came in with his face all blotchy...

CHLOE: ... with the vicar... basically just sits down and says:

CHLOE BARRY & CARLY: Breck's dead.

CARLY: Those were the words.

CHLOE: We broke down in tears.

CARLY: I remember us all on the couch just crying and crying... and holding onto each other in a state of shock.

SEBASTIAN: I don't remember crying. I was just, like, sitting there... just trying to... just racking my brain... so... you know... I don't feel like I had emotions pouring out... I was just kind of neutral.

CHLOE: I remember Dad telling us that basically Grandma and Grandad would be coming the next day, and Mum the day after that.

LORIN: The house was full of people I knew and police. We were constantly in police meetings, going over things.

LD was claiming Breck had killed himself. So, they had to go through that line of enquiry, which was so difficult... asking, you know,

POLICE/LORIN 2: Was he suffering from depression?

POLICE/LORIN 3: Was he sad?

POLICE/LORIN 2: Did he have friends?

LORIN: I just kept saying:

He was fine. We had arguments about the computer, but it was never, never anything deeper. He wasn't that kind of dark child.

LORIN: Detectives all dressed in white, ripped his room to shreds. It was very traumatic.

LORIN 4: LD immersed all the technology in water and scraped it with a screwdriver to ruin the evidence.

BARRY: Daynes knew it was Lorin's birthday and probably planned it for… he wanted it to be that day, to get back at her for… for taking away his internet plaything… whatever you call it… so, erm…

CARLY: We didn't get many details about what had happened.

CHLOE: All we knew was that Breck was dead. Grandma told me stuff but not everything. I remember at night, in my room, I was just looking it all up online…

CARLY: We probably shouldn't have, but we did.

CHLOE: People were messaging me saying, like, how sorry they were because everyone started finding out… and I was just, like, crying.

LORIN: The next morning we went to the hospital, hoping it wasn't actually him. He just looked so beautiful.

BARRY: He was in a little sheet and looked like a baby.

LORIN: I visited him in the morgue four times and made sure each visit was distinctly different.

The first, just blubbering that I was sorry for being such a horrible mom who hadn't saved him; the next, asking him why did he go there and why didn't he listen to me? Then told him stories about when he was a baby. Each week he was decomposing.

BARRY: I mean, his skin was falling off, it was ridiculous. No amount of make-up could cover it.

LORIN: The defence were dragging their feet about what tests they wanted to do.

BARRY: Daynes was being intransigent.

LORIN: Our team did all the tests straightaway and knew it wasn't a self-inflicted wound. We were begging them, begging them to release his body, to plan his funeral.

I wanted it on his birthday because I thought, it just seemed appropriate. We sang happy birthday at his funeral.

CARLY: We all wore blue to honour him.

LORIN: That next year was really hard, off work, and just trying to semi "be there" for the triplets. I only left the house for police meetings, and then have to be walked to the car. I couldn't judge distances or if it was safe.

The triplets not only lost a brother, but also a normal mom.

BARRY: I trusted people who took advantage of me, which is just unbelievable.

People say:

"You're so strong to be able get back and go to work."

My work suffered for years! It was a very traumatic time financially, and emotionally, and relationship wise and there's an irony here too… you don't want people to bring it up, but there's a point where, when they don't, it means it's being forgotten and that's kind of a shitty situation.

LORIN: My boyfriend, who I'm not with now, was a rock. He kept the triplets in normal life, where they just laughed and joked, whereas I couldn't.

BARRY: Just to end this whole train of thought is, like… I still have nightmares. I wake up at 3:33 almost every morning. I think that's when he must've died. It's bizarre and I'm always thinking about Breck.

LD *is unmasked by officials.*

LORIN: The first time I saw Lewis Daynes was via a video link in one of the hearings. He looked like a normal teenager, with a black T-shirt on, like something Breck would wear. I looked at him and thought, you could have been hanging out at our pool having pizza with us.

OLLIE: I was asked to appear as a witness and decided that sitting behind the camera was my preferred method. I didn't want to see Lewis. I felt it would bring up too much and I wouldn't be able to give a clear statement.

LORIN: When we got to the trial he'd aged and wouldn't look at me.

BARRY: He pled guilty.

LORIN: His mom visited him from Egypt, for the first time in years. Maybe she convinced him to plead guilty because there was so much evidence against him. Maybe it's why he'd pretended to be an Egyptian freedom fighter, bigging himself up after his mom abandoned him and moved there. In the sentencing, they go through a lot of the gory bits to show just how depraved he was.

LORIN 5: There was a mix of semen with wet blood. To mix and therefore to have been wet simultaneously, puts them close together in time.

LORIN 6: An opened box of condoms was found in a drawer. A splatter of blood indicates it hadn't been in the drawer when the blood had been splattered so, it gave them circumstantial evidence of a sexual element.

LORIN 5: They said Breck'd been given a "shocker stab", then tied up and gagged. We don't know for how long and then…

BARRY: The jugular vein is very… it was seconds… boom! The blood leaves your brain and, yeah, and I doubt that part of it was painful, but he was tied up and, we believe, raped. It had to be premeditated.

LORIN: When the judge said the murder was done in a sadistic and sexual manner his family sniggered. That's what he came from. They just made it worse.

BARRY: If my child had done that I'd have guilt beyond belief but, like I said at the outset, we should take a step back to see what caused Lewis Daynes to feel that this was something he needed to do.

LD: Hi there, I need police and a forensic team to my address please.

BARRY: I think his parents caused or exacerbated whatever he supposedly has… by not getting help for him or giving him the care he needed.

LD: My friend and I got into an altercation and I'm the only one who came out alive.

BARRY: His mother reportedly met some guy when she was on holiday in Egypt, married him and never came back, so for her to actually show up was surprising.

LD: I, in self defence, put my left arm up to block him from stabbing me effectively… can you… can you not interrupt me on this part…

BARRY: He'd spent a whole year in custody. She hadn't been to see him. Nobody came to see him. Nobody!

LD: *(Deep breath.)* I grabbed a knife and I stabbed him once in the back of the neck, I believe somewhere near the brainstem.

LORIN 5: There was one point where, in prison, he tried to say that terrorists made him do it. His whole life was built around lies, manipulation and control.

LD: He turned around to try to carry it on. I think I stumbled on my chest of drawers. I fell over.

LORIN: My heart breaks over and over wondering how long Breck was scared and in pain.

LD: I don't remember exactly what happened, but the fight ended with me cutting his throat.

BARRY: The moment he realised that what we were saying was true and this guy was bad, must have been as scary as hell and I don't know how long that lasted. Hours? A whole day?

LD: I believe I turned around and slashed his throat.

LORIN: He got a life sentence, with a minimum of 25 years.

LORIN 5: A psychologist said he did this because it's what makes him feel alive… what feeds him. He'll reflect on it, engage with it in his memories. That's really terrifying.

LORIN: We… we… two weeks after Breck was killed, I said:

Something good has to come from this as Breck was innately good. I want to set up a charity, so people can learn about grooming and the dangers online. It was tough, but I did it.

BARRY: I've got three other children and they don't deserve to be, er, you know, side-lined, or Breck put on top of them. I also have a life to lead. You know, this has happened and we all have to live with that shadow.

LORIN: I guess my goal is for this play to tell Breck's story without my being tied to doing so forever, because it's really hard telling Breck's tragic story over and over again.

BRECK: You don't have to do this.

*The triplets enter to offer **LORIN** support.*

CARLY, CHLOE & SEBASTIAN: Oh, Mum!

CHLOE: We love you

CARLY: Don't cry.

CARLY: We'll get back to ourselves.

CHLOE: We just need time.

LORIN: I kind of just wannabe normal again one day and it's really… it's like I don't have that option.

LORIN 4: I don't feel whole…

LORIN 2 &3: … missing parts…

LORIN 4 & 5: … and can't function properly.

LORIN 1-5: Torn apart.

LORIN: Sometimes I hear Breck say:

BRECK: I know it's hard… but Mom, you've nothing to feel guilty about… I just want you to get better.

*(**LORINS** 2-5 are gently shown offstage by the triplets, who remain with **BARRY**, looking on.)*

LORIN: When poor Breck was killed, I died too. I couldn't leave the house or face people or noises very easily. My boyfriend'd take me and the triplets out to the sea, or the woods, and I'd just sit and watch the kids play and look at nature in wonder. The blue sky, there's Breck, the beautiful sea, that's Breck, falling leaves, is that Breck? Or a unique stone so I could pick up and 'hold' my Breck. I was trying to find him, to keep him with me through my bond with nature. All through these years I have collected stones, especially heart-shaped stones, like calming or comforting charms, some sort of proof that the wonder of Breck is still with me.

BRECK *goes to one of the heart-shaped rocks and picks one up.*

BRECK: Mom… I am still with you and Dad and the triplets.

*(**BRECK** hands the stone to **LORIN**.)*

See? I'll always be with you.
All of you…
… always.

LIGHTS FADE.

Digital Resources for Teachers

There are a number of practical digital resources for teachers and students who are studying *Game Over* as a set text.

Please see the *Game Over* page on www.salamanderstreet.com for further details.

You can see Mark talking about the key moments in the life of *Game Over* on his Mark Wheeller Youtube Channel in his Wheellerplays Series Episodes 88-89

Teachers – if you are interested in buying a set of texts for your class please email info@salamanderstreet.com – we would be happy to discuss discounts and keep you up to date with our latest publications and study guides.

Missing Dan Nolan
Paperback 9781913630287
eBook 9781913630294

Chicken!
Paperback 9781913630331
eBook 9781913630324

Chequered Flags to Chequered Futures
Paperback 9781913630355
eBook 9781913630348

Hard to Swallow
Paperback 9781913630249
eBook 9781913630256

Too Much Punch For Judy
Paperback 9781913630300
eBook 9781913630317

Hard to Swallow, Easy to Digest
(with Karen Latto)
Paperback 9781913630409
eBook 9781913630393

Hard to Swallow, Easy to Digest: Student Workbook
Paperback 9781913630416
eBook 9781913630423

The Story Behind … Too Much Punch for Judy
Paperback 9781913630379
eBook 9781913630386

Salamander Street will be publishing new editions of Mark's plays in 2020 – follow us on Twitter or Facebook or visit our website for the latest news.

www.salamanderstreet.com